Is this
an
Agate?

An Illustrated Guide
to
Lake Superior's Beach Stones
Michigan

**Written and Illustrated
by
Susan Robinson**

Table of Contents

Introduction 1

Information for Collectors 1

Geological History of the Keweenaw Region 2

Mineral Descriptions 5

    Quartz 5

    Chert 5

    Agate 6

    Jasper 7

    Chlorite 7

    Epidote 8

    Prehnite 8

    Calcite 9

Minerals Rarely Found 9

    Copper 10

    Datolite 10

    Chlorastrolite (Greenstone) 10

    Thomsonite 10

Rock Descriptions 11

    Basalts 11

    Gabbro 12

    Porphyry 13

    Rhyolite 13

    Granite 14

    Gneiss 14

    Slate 15

    Sandstone 15

    Conglomerate 16

    Shale 16

    Limestone 17

Museums 17

Maps and Directions to the Beaches 18

Glossary 22

Acknowledgements 23

Introduction

Is there anything more relaxing than strolling along a beach, surveying the multitude of pebbles, and listening to the waves? The region covered in this little book abounds with open beaches, enticing people to walk along the shore and enjoy the beauty of Lake Superior and its fascinating pebbles.

This handy take along guide will help you identify these pebbles and thus remove some of their mystery. While it discusses the most common rocks and minerals found on these beaches, the text does not cover every exotic rock type that may be encountered. The geological history of the south shore of Lake Superior is ancient and complex; most of the rocks you see are over a billion years old. Many of the pebbles found on the beaches were transported here by glaciers, and tumbled smooth by the lake's waves.

I hope this book is helpful for everyone who enjoys walking along Lake Superior's beaches as much as I do. Have fun beachcombing!

Information for Collectors

This book describes both rocks and minerals. **Minerals** are naturally occurring solid substances formed by geological processes. They have definite chemical compositions and crystal structures. **Rocks** are made up of various individual minerals. In some rocks the individual mineral grains can only be seen under a microscope; in others they are large enough to be identified by sight.

The quartz family of minerals will be discussed first, followed by other minerals arranged by their color. Descriptions of the various rocks found on the beaches are given next, arranged by their origins as igneous, metamorphic or sedimentary. A note on the relative abundance of the rocks and minerals is given at the end of the Best Beaches information. These terms are explained below:

Abundant	=	easily found
Common	=	usually present
Uncommon	=	occasionally found
Rare	=	not usually found
Very Rare	=	almost never found

Note: All of the illustrations show the way the rocks and minerals appear when dry, unless noted otherwise.

Geological History of the Keweenaw Region

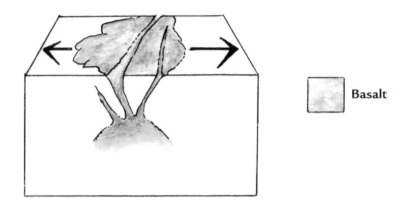

Basalt

1) Approximately 1 billion years ago, a hot spot in the Earth's mantle caused a rift valley to form in the center of the North American continent. Molten rock found its way through faults and fissures to the surface, where it cooled, and formed basalt. Escaping gas bubbles were trapped near the tops of the cooling lava flows, forming vesicles in the basalt.

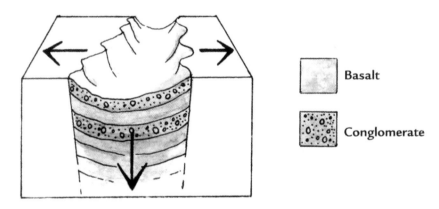

Basalt

Conglomerate

2) During quiet periods of volcanism, weathering and erosion of the surrounding rocks covered the basalts with sediments that later formed conglomerate. As more rifting took place, lava again poured out, covering the conglomerate. This sequence was repeated many times. Eventually, the continued rifting and weight of accumulated rock caused the whole region to slowly sink.

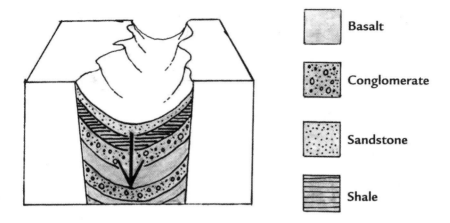

	Basalt
	Conglomerate
	Sandstone
	Shale

3) The volcanic eruptions finally ended, but continued erosion caused most of the region to be covered by more conglomerate, then shale and sandstones.

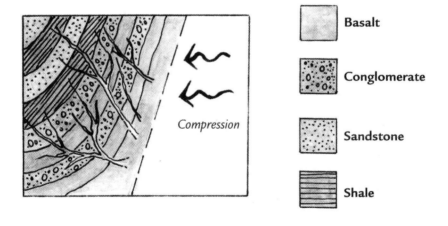

	Basalt
	Conglomerate
	Sandstone
	Shale

Compression

4) A continental collision from the east caused the entire sequence of alternating igneous and sedimentary rocks to be tilted, fractured, and faulted. Hot water containing dissolved copper and other minerals moved upward from deep within the rift and invaded open spaces in the rocks. Thus, copper and other minerals were deposited in the fractures, porous conglomerates, and vesicles in the tops of the lava flows (forming amygdaloidal basalts).

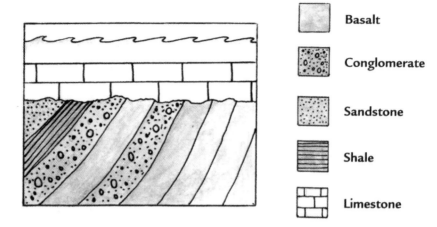

5) Over the next several hundred million years, encroaching seas gradually covered the area, depositing marine sediments that would eventually form limestone. When the seas retreated, the rocks were again exposed to weathering and erosion.

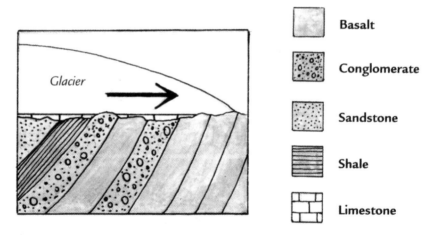

6) About 3 million years ago, a climate change saw the onset of the Ice Age. Massive glaciers (up to a mile thick) moved southward, eroding the soft limestone and other sedimentary rocks, and exposed the underlying basalts, which contained copper, agate, and other minerals. These were carried by the glaciers and deposited across several states in the upper Midwest.

About 10,000 years ago, the last glaciers eventually retreated, leaving behind Lake Superior and our present landscape.

Mineral Descriptions

The Quartz Family

Consisting of silicon dioxide, quartz is one of the most common minerals in the world. Many forms of quartz are coarsely crystallized, but most of the varieties presented in this booklet are not. All are hard and easily scratch glass.

Quartz   Pebbles of pure quartz are probably weathered out from veins in rock. Most of the larger quartz pebbles found on the beaches along the western shoreline have probably been transported here by glaciers, and did not originate in this region. <u>Color:</u> Any shade of white, gray, or light yellow, depending on the amount and type of impurity. <u>What to look for:</u> Quartz pebbles most nearly resemble white, frosted glass, and are usually translucent. <u>Best Beaches:</u> Any of the beaches listed. **Common.**

Chert   Like agate, this variety of quartz is usually opaque. It can sometimes show crude banding, but not as regular and distinct as the banding in agate. <u>Color:</u> Any range from gray, tan, brown, and yellowish. <u>What to look for:</u> Chert has a waxy, sometimes shiny appearance from being polished by wave action. <u>Best Beaches:</u> Most beaches listed contain pebbles of chert, but chert is less common on the eastern side of the Keweenaw Peninsula. **Common.**

Agate   Easily recognized by its bands of varying colors, agate is a common variety of quartz. Lake Superior agates can range from being completely opaque to translucent. They rarely show transparency. They may range in size from a pea to a grapefruit. <u>Color:</u> The bands range from white, bluish-white, or gray, to brown, reddish-brown, tan, orange, and shades of yellow. Red, yellow, and orange bands contain varying amounts of iron, causing their colors. <u>What to look for:</u> Scan the beach for pebbles that display a waxy, sometimes pitted orange peel appearance, and are brown, tan, pinkish, or gray in color. The translucency may not be readily obvious, unless the agate is wet. The pitted outer skin is evidence of their formation in vesicles in the basalt. Most agates found today are under an inch. <u>Best Beaches:</u> Agates can be found on any beach, since they have been randomly dispersed by the action of waves and glaciers. However, the best places are from Little Girl's Point to Keweenaw Point, and from Grand Marais to Whitefish Point. **Uncommon to rare.**

in amygdules

Jasper This is a close cousin of chert, but is much more colorful. It, too, is always opaque due to impurities. Bright red jasper is common in many of the iron formations across Michigan's Upper Peninsula and northern Wisconsin, where it occurs as bands in the iron ore (jaspilite). Color: Jasper can range from red to yellow or brown, and any color in between. The colors are caused by iron impurities. What to look for: Jasper's bright colors make it easily recognizable. It has the same surface appearance as chert. Jasper is sometimes associated with silvery black metallic-appearing hematite. Best Beaches: Jasper may be found on most of the beaches listed, but is less common on the eastern side of the Keweenaw Peninsula. Red pebbles, probably eroded from the iron formations, occur at Little Girl's Point. **Uncommon.**

Green minerals

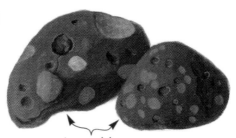

in amygdules

Chlorite This is a common vesicle-filling mineral in the Keweenaw basalts that is often confused with chlorastrolite, due to its similar color. Chlorite is softer than chlorastrolite, and has a platy to granular appearance on broken surfaces, whereas chlorastrolite has a more fibrous to radial structure. Color: Dark green to nearly black. What to look for: Chlorite is most often found filling amygdules in basalt. Chlorite amygdules that are freed from rock by weathering seldom last long due to their softness. Best Beaches: Any of the beaches along the western side of the Keweenaw Peninsula. **Common in basalt.**

in flow-banded rhyolite

in amygdules

Epidote   A relatively common mineral in the Keweenaw, epidote is found in veins and amygdules in the basalts. It frequently occurs with quartz. <u>Color:</u> Pale yellow-green to dark green, and usually appears opaque. <u>What to look for:</u> The chartreuse green color of epidote is distinctive. Some of the veins and amygdules may not be completely filled, and contain only a crystallized crust of epidote. Epidote also occurs as small pebbles free from the enclosing basalt, and in the variety of altered granite known as unakite. <u>Best Beaches:</u> Tamarack Waterworks, Calumet Waterworks, Agate Beach (near Toivola), and McLain State Park. **Common.**

in amygdules

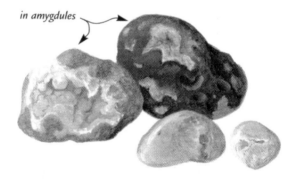

Prehnite   Like epidote, prehnite is another green mineral found in the basalts of the Keweenaw. Prehnite fills cracks and veins and sometimes forms hilly, compact masses in partially filled crevices or amygdules. <u>Color:</u> Ranges from white to yellow-green, to a pale mint green. Prehnite also occurs on some beaches as attractive wave-polished pebbles, displaying a pretty combination of radiating pale green, white, and pink colors. These are often mistaken for thomsonite, which is rarely found.   <u>What to look for:</u> Prehnite has a pearly to waxy luster and translucent appearance, especially when wet. In contrast to epidote, prehnite is a much paler, cooler green color. <u>Best Beaches:</u> Western side of the Keweenaw Peninsula, from Calumet Waterworks beach to Keweenaw Point. **Common.**

White minerals (other than quartz)

Calcite A very common mineral that consists of calcium carbonate. Calcite can easily be scratched with a knife blade; quartz cannot. <u>Color:</u> Calcite can be many colors, but white to pale yellow are the most common. <u>What to look for:</u> Calcite varies from opaque to translucent, to (rarely) transparent. It nearly always displays some flat cleavage planes on its surface due to its internal crystal structure. Calcite commonly fills cracks and vesicles in basalt. Because it is so soft, calcite weathers much faster than other minerals, so it is uncommon as pebbles. <u>Best Beaches:</u> From Calumet Waterworks to Great Sand Bay. Common as amygdule fillings, uncommon otherwise.

Minerals rarely found on the beaches today

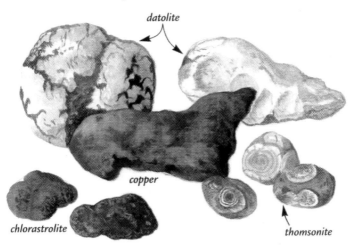

datolite

copper

chlorastrolite

thomsonite

The four minerals shown above and explained on the next page were all formed within the basalts, and may occasionally be found along the Keweenaw Peninsula's western shoreline. The chance of spotting any of them on the beaches today is rare. They are more commonly encountered in the mine dumps of the former copper mines.

Copper This mineral is the "red metal" for which the Keweenaw is world-renowned. Glacially transported pieces of copper are known as "float copper." Copper is usually found as irregularly shaped, rounded masses from 1 to 3 inches, though huge masses are known. Its true, bright reddish-orange color is often masked by dark brownish-red or green-blue colors, due to oxidation. The copper found in basalt at Great Sand Bay was transported there from mine dumps to help prevent beach erosion.

Datolite The colored, porcelain-like masses of datolite found in this region are unique. Water-worn nodules of datolite have chalky white exteriors, though their interior colors can range from white to yellow, pink, peach, and shades of red, all dependent on the amount and type of impurities. Datolite rarely has yellow, green or bluish colors.

Chlorastrolite (Greenstone) Chlorastrolite (or "Isle Royale" greenstone) is Michigan's state gemstone. The correct mineral name for chlorastrolite is pumpellyite. Greenstones consist of tightly compact radiating masses of pumpellyite crystals that, when polished, display a crackled surface appearance. Greenstone pebbles over an inch are rare; most are peasized or smaller, and have a powdery light green to dark green surface color. Most greenstone found along shorelines today is at Isle Royale National Park, but collecting there is prohibited*.

Thomsonite Only a few basalt flows in the Keweenaw contain this mineral, which forms in tight, radiating masses that fill vesicles. These form distinctive "eyes" displaying circular bands of alternating color combinations of black, green, pink and orange, on a creamy white background. Thomsonite pebbles are usually under 1/2 inch in size. Small pebbles of radiating green and pink prehnite are much more common, and easily mistaken for thomsonite.

* **Information on collecting on Isle Royale National Park:** Since this island is a federally regulated National Park, collecting agates, greenstones, and datolite from both the shore and underwater is not allowed. You may photograph or sketch the specimens you find, but please leave them on the island and do not remove them from there.

Rock Descriptions

Igneous rocks are those formed when molten rock, or magma, solidifies. Magma that reaches the Earth's surface is known as lava.

Basalt A dark-colored volcanic rock whose individual minerals are too small to be identified by the naked eye. Basaltic lava is extruded on the Earth's surface and cools quickly, thus causing the fine-grained texture. Basalt makes up a high percentage of the bedrock in the Keweenaw Peninsula and Isle Royale. Four different varieties are recognized, and discussed below. Color: Dark greenish gray to black, sometimes dark reddish brown. What to look for: Pebbles of any size in the colors and textures noted. Best Beaches: Basalt is found on most of the beaches in this guide, but is less common along the eastern side of the Keweenaw Peninsula. **Very abundant.**

Massive basalt This is just what it is named, a solid mass of basalt – no holes or crystals visible in the rock (see above illustration).

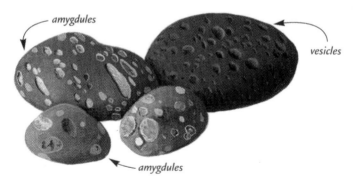

amygdules

vesicles

amygdules

Vesicular basalt A basalt which is full of empty holes or vesicles, the evidence of gas bubbles in the lava.

Amygdaloidal basalt A vesicular basalt whose vesicles have been filled in with minerals.

Ophitic basalt (diabase) This basalt contains areas of indistinct, lighter colored, radiating crystals of feldspar (which resemble small, out-of-focus dandelion heads gone to seed) in a dark groundmass. Ophitic texture usually can be seen more easily when pebbles are dry.

Gabbro

Gabbro A coarse grained igneous rock with a texture like granite, but nearly the same composition as basalt. <u>Color:</u> Dark overall; gabbro appears dark gray, black to greenish gray. <u>What to look for:</u> Dark rocks that have a very coarse grain. <u>Best Beaches:</u> Most beaches in this guide will yield some gabbro, but the best ones are from Ontonagon to Little Girl's Point, and the Whitefish Point area. **Uncommon.**

granite porphyry

basalt porphyry

Porphyry An igneous rock that contains large, distinct crystals dispersed throughout a finer-grained groundmass. Basalt, rhyolite and granite can all display porphyritic texture. <u>Color:</u> Porphyry can be almost any color, since a variety of igneous rocks can have this texture. <u>What to look for:</u> In the Keweenaw, many of the basalts, rhyolites, and granites are porphyritic. Look for light-colored, rectangular to blocky-shaped crystals scattered throughout a darker, fine grained rock. Some porphyries may resemble amygdaloidal basalt, but **amygdules are round, not angular or square.** <u>Best Beaches:</u> Porphyry can be found on any of the beaches in this guide. **Common.**

flow bands

Rhyolite   Like basalt, rhyolite is another fine-grained volcanic rock that often may contain vesicles and amygdules. However, its composition is the same as granite, and therefore it has an overall lighter color than basalt. <u>Color:</u> Ranges from reddish brown to pinkish tan. <u>What to look for:</u> Any dense, fine-grained reddish colored rock. Many rhyolites found on the beaches are flow-banded and resemble jasper, but do not have its color intensity. <u>Best Beaches:</u> Can be found on most beaches, but is less common on the eastern side of the Keweenaw Peninsula.   Rhyolite is particularly abundant at Copper Harbor, since it comprises many of the pebbles in the Copper Harbor Conglomerate. **Abundant.**

green epidote

unakite

Granite An igneous rock that consists mainly of the minerals quartz, feldspar and mica. Its grain size can be from medium (a salt and pepper look) to coarse (grains an inch or more across). Some granites may develop an exceptionally coarse grained phase known as a granitic pegmatite. Most of the granite pebbles found in this region were transported from Canada by glaciers. Color: Depending on the color of the feldspar, granites can range from white, gray, pink, to even reddish, all with black specks scattered throughout. What to look for: Granite is readily obvious because of its coarse grain and speckled appearance. Altered granite containing epidote is called unakite by lapidaries, and has a distinctive orange and green color combination. It is also sometimes found on the beaches. Best Beaches: Any of the beaches in this guide. **Abundant.**

Metamorphic rocks are formed by heat and pressure acting on other previously formed rocks. This causes changes in both the texture and mineral composition of the pre-existing rock. Metamorphic rocks often show a banded or layered structure.

Gneiss Most gneiss is similar in composition and appearance to granite, but shows a pronounced alignment of its mineral grains into bands. It can be fine to coarse-grained. Color: The colors range (even within the pebble) from white, pink, and greenish to light gray. What to look for: Light-colored pebbles displaying alternating light and dark-colored bands of minerals. Occasionally, gneiss will contain small, reddish brown garnet crystals, usually a 1/4 inch or less in size. Best Beaches: Any of the beaches in this guide. **Abundant.**

Slate Slate is a metamorphic rock that forms when shale is subjected to heat and pressure. Some of the so-called slate found at Keweenaw Bay is actually phyllite, a rock containing more mica and formed under greater heat and pressure. <u>Color:</u> Dark to light gray. <u>What to look for:</u> Flat, smooth pebbles displaying some layering. Pebbles of slate (and especially phyllite) that are dry will usually exhibit a shiny, satiny surface. <u>Best Beaches:</u> Common from Keweenaw Bay beach northward along the eastern shore of Keweenaw Bay; **uncommon elsewhere.**

Sedimentary rocks are generally formed in two ways. Some, such as sandstone, shale or conglomerate, form when sediments (e.g., sand, clay, or pebbles) are compacted and cemented together under pressure. Others, like limestone, form by chemical precipitation from lakes and oceans.

Sandstone A sedimentary rock consisting mainly of sand-sized grains of quartz cemented together. <u>Color:</u> From white, beige, to light tan, gray, and even deep reddish-orange, depending on the amount of iron oxide present. <u>What to look for:</u> Flat rocks showing layering and having a coarse, gritty texture and feel. Some larger pieces may show preserved ripple marks and mud cracks. <u>Best Beaches:</u> Any beach along the eastern side of the Keweenaw Peninsula from the South Portage entry (White City) to Lac La Belle. Other beaches where sandstone is abundant include Saxon Harbor, Little Girl's Point, Tamarack Waterworks, McLain State Park, and the beaches between Copper Harbor and Esrey Park. Pictured Rocks National Lakeshore is made up entirely of sandstone, but collecting there is prohibited. **Abundant.**

Conglomerate An easily recognized sedimentary rock consisting of pebbles cemented together by finer particles. The pebbles can range from tiny to grapefruit size. Most of the pebbles in the conglomerates of the Keweenaw Peninsula are rhyolite. <u>Color:</u> Conglomerate typically ranges from dark reddish brown to burgundy, to an orange brown, due to iron oxide. Angular pebbles in a whitish gray, sandy matrix is probably concrete from industrial sites. <u>What to look for:</u> The distinctive texture of pebbles cemented together is unmistakable. <u>Best Beaches:</u> Most beaches along the western side of the Keweenaw Peninsula. **Common to uncommon.**

Shale A very fine-grained sedimentary rock that formed by the compaction of a clay- or mud-like sediment. <u>Color:</u> Shales in this region are usually dark-colored, in various shades of gray. <u>What to look for:</u> Flat, thin rocks. Some may exhibit layering. Most make excellent skipping stones. <u>Best Beaches:</u> Any beach from Ontonagon through the Porcupine Mountains Wilderness State Park, where it is locally common to abundant. **Uncommon to rare elsewhere in the area.**

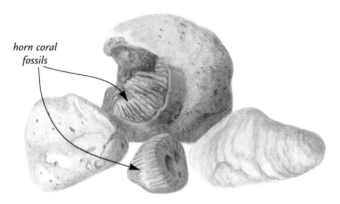

horn coral
fossils

Limestone A sedimentary rock made up of calcium carbonate. Many limestones contain fossils, evidence of plants and animals from the distant past. The most common fossils found in limestone in this region are corals and shells from the Ordovician period, about 400 million years ago. Color: Extremely variable, but usually gray, brown or tan to nearly white. What to look for: Dull, light colored rocks showing some layering or fossils. Chert can look similar, but is much harder and slightly glossy. Best Beaches: Any of the beaches listed in this guide. **Common to uncommon.**

Mineral Museums you can visit

The A. E. Seaman Mineral Museum: Located on the fifth floor of the Electrical Energy Resources Building on the Michigan Technological University campus in Houghton, Michigan. Besides the best collection of Lake Superior region minerals, the museum also has extensive collections of worldwide minerals. Parking: At the visitor's lot off highway 41 across from the Public Safety office, where you may obtain a parking permit. Metered parking is also available across campus. Visiting Times: Mid-May to November: M-F 9 a.m. to 4:30 p.m., Saturday 12-4 p.m., closed Sunday. November to Mid-May: M-F 9 a.m. to 4:30 p.m., closed weekends. For information or to confirm hours, phone (906) 487-2572. Website: http://www.geo.mtu.edu/museum/

Gitche Gumee Agate and History Museum: Located at E21739 Braziel St., Grand Marais, Michigan; owned and curated by Karen Brzys. The museum displays Lake Superior agates as well as minerals from various localities. It also has many historical artifacts from the area. There is adequate parking anywhere along the street. Visiting times: Open from June to October. For additional information or to schedule tours, phone: (906) 494-2590 or FAX (906) 494-2480.

Directions to Beaches

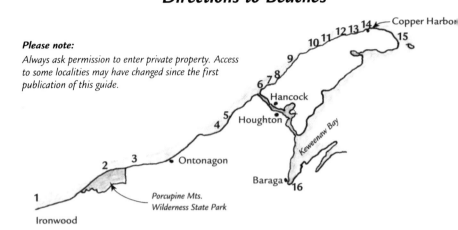

From Ironwood to Houghton

1   Little Girl's Point

From Highway 28 in Ironwood, turn north onto Lake Street (County
Highway #505). There is a traffic light at the intersection. Follow #505
approximately 17 miles to the point. Follow #505 another 9 miles west,
and turn north onto a dirt road (signed) to access Saxon Harbor, where
there is another extensive beach.

2   Porcupine Mountains Wilderness State Park

From the intersection of Highways #64 and #107 in Silver City, proceed
west on #107 for 3 miles to the park's entrance. Follow the signs to the
Visitor's Center, where you can obtain directions to several good beaches.

3   Ontonagon area

From the village of Ontonagon, proceed west on Highway #64 across the
bridge. Green's Park Beach is 6 miles from the bridge. There are other
public access points at 11.4 miles and at 12.5 miles, which is where the Big
Iron River empties into Lake Superior.

4   Misery Bay Beach

Proceed south from Houghton on M-26 (nearly 11 miles) to Toivola. Turn
right onto Misery Bay Road and drive 6 miles to where the road to Misery
Bay forks left. Continue 5 miles to the parking lot for the beach.

5   Agate Beach

Follow the directions for Misery Bay to where the Misery Bay Road forks
left. At this point, turn right for Agate Beach and proceed for 1 mile. Turn
right again and drive 2 miles to a "T" intersection. Turn left and drive 0.7
mile to the Agate Beach parking area.

The Keweenaw Peninsula (west side)

6 McLain State Park

McLain State Park is on Highway #203 midway between Calumet and Hancock. From the intersection of Highways #41 and #203 in Calumet, take Pine Street (#203) south for 8.5 miles to the park entrance on the right. From the intersection of Highways #203 and #41 in Hancock, follow #203 (Quincy Street) for 9.5 miles to the park entrance on the left.

7 Calumet Waterworks Park Beach

From the intersection of Highways # 41 and #203 (Pine Street) in Calumet, take Highway #203 2.6 miles to the Calumet Waterworks Road (Lake Shore Drive). Turn right and drive 2.5 miles to the park entrance on the right.

8 Tamarack Waterworks Beach

From the intersection of Highways # 41 and #203 (Pine Street) in Calumet, take #203 1.4 miles and turn right onto Tamarack Waterworks Road. Take the Waterworks Road for 1.6 miles and turn left at the fork. Continue for 1.2 miles to Sedar Bay Road, where you turn left again. Proceed 0.8 miles to the parking area for the beach. The road dead-ends here, and is bordered by private property. Please ask permission.

9 Sunset Bay Beach

From the intersection of Michigan State Highway #26 and the road to Fivemile Point, just south of Eagle River, take Fivemile Point Road for 4.7 miles, and turn right on the dirt road to Sunset Bay Campground. After crossing a tiny wooden bridge, proceed straight ahead for the campground. You may collect here for a fee of $5.00. The road to the left (after the bridge) leads to Seven Mile Point. Access is limited to weekends from Memorial Day to Labor Day. For information contact: www.northwoodsconservancy.org

10 Eagle River

Access to beaches at Eagle River is easy. Turn north off Highway #26 after crossing the bridge (going east) or just before the bridge (going west). Drive to the Lake's shore and park.

11 Great Sand Bay

Approximately 14 miles west of Copper Harbor on Highway #26 is Great Sand Bay. This is one of the largest beaches in the Keweenaw; it has a great view, a nice sand beach, and an extensive pebble beach for well over a mile.

12   Esrey Park

On Highway #26, 4.4 miles west of the entrance road to Copper Harbor
Marina is a small public park. The beach here is almost non-existent, and
consists mostly of large boulders of basalt.

13   Conglomerate and sandstone exposures

Along Highway #26, between 1.9 and 3.3 miles west of the entrance road
to Copper Harbor Marina, there are good exposures of conglomerate and
sandstone along the shore. These rocks were tilted upward from their original horizontal position about a billion years ago. Note the good examples
of cross-bedding in the sandstone exposures.

14   Copper Harbor (Access may be restricted)

Drive west on Highway #26 from Copper Harbor to the Copper Harbor
Marina sign. Turn right and park. To the left is a sign for Hunter's Point
Trail. Follow the trail around the harbor's edge to the beach.

15   High Rock and Keystone Bays

These beaches are located at the very end of the Keweenaw Peninsula at
Keweenaw Point. The dirt roads that lead to them require a 4WD high
clearance vehicle. It is best to ask for directions to them in Copper Harbor,
as the roads frequently branch out and are unsigned.

The Keweenaw Peninsula (east side)

There are several beaches on the eastern side of the Keweenaw Peninsula,
but they consist almost entirely of sand and sandstone.

16   Keweenaw Bay Beach

From the intersection of Highways #41 and #38 in Baraga, take Highway
#41 south for 2.4 miles, to where it turns east around the end of the bay.
There is adequate parking along the left shoulder near the sand beach.

Marquette to Whitefish Point

Most of the accessible beaches from Marquette eastward through the
Pictured Rocks National Lakeshore consist of sand and sandstone. Beach
collecting within the boundary of the National Lakeshore at Pictured Rocks
is prohibited.

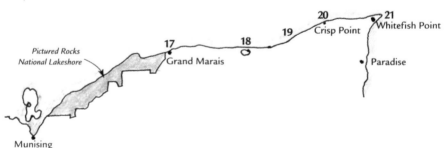

17   Grand Marais area

From within the village of Grand Marais, follow Highway #77 to its end, and turn west onto Braziel Street, then proceed 0.3 mile to Woodland Park Beach. At the far west end is a wooden stairway to the beach. About 100 yards to the west is the Pictured Rocks National Lakeshore boundary, marked with a sign. You may collect anywhere eastward of that sign.

Agate Beach at Grand marais can be accessed by following Lake Avenue west from main Street for 0.2 mile.

Old East Bay beach can be accessed by following County Highway #58 east about 1.8 miles from Grand Marais, and turning left at the Lutheran Cemetery, near where the Sucker river empties into Lake Superior. Go about 0.2 mile to the beach.

Superior Dunes beach is about 9.6 miles east of Grand Marais on highway #58 (#58 and #407 run together here). You will have to park along the road and walk north up the hill and over the dunes about 0.3 mile to the beach.

Lake Superior State Forest Campground is more easily accessed, and is 2.4 miles further east along Highways #58 and #407.

18   Muskallonge Lake State Park

From Lake Superior State Campground proceed east on #58/407 about 7 miles to Muskallonge Lake State Park. Access to the beach is across the road from the campground.

19   Mouth of the Two-Hearted River

Continue on Highways #407 and #H37 southward from Muskallonge Lake, then turn left (east) onto Highway #410. Continue until #410 joins #412, and drive east on #412 to its intersection with #423. Turn left on #423 and drive north until you reach the beach. This area can also be accessed by following County Highway #500 northward off of Highway #123. Beach access in this area may require a 4WD vehicle at certain times of the year.

20   Crisp Point

Crisp Point lies at the end of County Highway #412, approximately 8miles from its intersection with County Highway #500. Access to this area may require a 4WD vehicle at certain times of the year.

21   Whitefish Point

From Highway #123 in Paradise, follow Whitefish Point Road 10 miles north along the shore to the point. There is a large parking lot and several accesses to the beach. Please obey he signs, since this area is a wildlife refuge and renowned birding area.

Glossary

Amygdule: (ah-mig'-dool) A gas bubble in volcanic rock that has been filled in with minerals.

Chlorastrolite: (klor-as'-tro-lite) A gemological synonym for green, radiating pumpellyite (greenstone).

Cross-bedding: A sequence of thin beds inclined at an angle to the main bedding plane of a granular sedimentary rock. Cross-beds signify the presence and direction of currents (wind or water currents) that deposited the original sediment.

Crystal: The external planar form or shape mineral assumes due to its internal ordered arrangement of atoms.

Greenstone: A local name synonymous with chlorastrolite.

Hematite: The major ore mineral for iron. It consists of iron oxide.

Igneous: The class of rocks formed when lava or magma cools and solidifies.

Lava: Magma that reaches the Earth's surface.

Mantle: The region of the Earth between its crust and outer core.

Metamorphic: A rock whose composition and/or texture has been changed by heat and pressure.

Opaque: The non-transmission of light through a substance.

Oxidize: To react with oxygen to form new minerals or compounds.

Porphyry: (pore'-fur-ree) An igneous rock with a texture characterized by individual, large crystals in a fine-grained groundmass or matrix.

Rift: An area in the Earth's crust that is spreading apart. The opening usually allows magma to come to the surface.

Sedimentary: Rocks formed from the deposition of mineral fragments or precipitates.

Translucent: The partial transmission of light through a substance

Vesicles: (ve'-sik-kills) Small, rounded holes in volcanic rocks caused by gas bubbles trapped as lava cools.

Volcanic: Any region, rocks or minerals that are derived from or influenced by volcanoes.